MATH FACTS

SURVIVAL GUIDE TO BASIC MATHEMATICS

THEODORE JOHN SZYMANSKI

TOMPKINS CORTLAND COMMUNITY COLLEGE

WADSWORTH PUBLISHING COMPANY
BELMONT, CALIFORNIA
A DIVISION OF WADSWORTH, INC.

P9-BID-698

Mathematics Editor: Anne Scanlan-Rohrer
Print Buyer: Barbara Britton
Signing Representative: Dawn Burnam
Printer: Malloy Lithographing

2 3 4 5 6 7 8 9 10—95 94 93 92

ISBN 0-534-17154-0

MATH FACTS was written for use in developmental mathematics courses to fill a need expressed by students for "anchor points" in basic mathematics. The guide began as an index card project that provided minimal but essential information on single mathematical concepts. A sixteen semester collection of cards has resulted in MATH FACTS.

The alphabetical listing offers the student entry points into basic concepts and math terminology. Each page is a simple treatment of a single concept. Reference is made at the bottom of each page to related ideas. The language chosen is that of operational definitions that have proven to be successful with developmental and returning students, particularly those with holistic learning styles. Cross-references are made to terms that may be easily confused.

Students find that MATH FACTS reduces anxiety by its high "safety level," i.e., they feel comfortable with its compact size and lack of intimidation.

MATH FACTS is designed as a stand-alone shelf reference and for use with a textbook or workbook in basic math courses. MATH FACTS is also being used by students enrolled in other courses where a brief review of basic math concepts is helpful such as introductory algebra, biology, nursing math and technical math.

TABLE OF CONTENTS

TOPICS IN THE GUIDE ARE ARRANGED IN ALPHABETICAL ORDER

ABSOLUTE VALUE

Symbol | |

Description: A way to indicate that the value of any quantity is positive, regardless of its sign.

Uses: Distances are always positive, but may come out negative in calculations, so one always "takes the absolute value" of a distance.

Example: The absolute value of -5 = 5

$$|-5| = 5$$

Rule: Evaluate the expression within the | | sign, following the rules of the *order of operations* and make that quantity positive.

Example: $|6 - 10 - 8| = 12$

Absolute values are used often in the physical sciences when expressing changes in measurements where the change is important and not the direction.

AREA

An area has two dimensions, length and width. It is the measure of flat objects such as carpets, walls, surfaces. The units of measure of area are *square* units such as *square* feet, *square* yards, *square* miles.

The area of almost any figure can be calculated by multiplying its length times its width.

Area of a square = length x width (length = width!)
Area of a rectangle = length x width
Area of a triangle = half the length (height) x width (base)
Area of a circle = pi x radius squared
Area of a trapezoid = average of widths x height

AREA is different from *PERIMETER*. The peRIMeter is the distance around the RIM of the object. Perimeter is a LENGTH, a one-dimensional measure. It is measured in inches, feet, miles, which are one-dimensional units.

3

CONTINUED NEXT PAGE

EXAMPLE: Calculate the area and perimeter of a garden that is 18' x 24'

AREA = L x W

AREA = 24' x 18' = 432 sq ft

PERIMETER = L + L + W + W

PERIMETER = 24' + 24' + 18' + 18' = 84'

24'

18'

AREA OF A CIRCLE

FORMULA: AREA = PI X R^2

See CIRCLES - TERMS - AREA

AREA OF A RECTANGLE

The area of a rectangle is given by L x W. Simply multiply the length by the width. Note that the units become squared. 3 feet times 7 feet = 21 square feet. The units are important.

Example: Calculate the area of a rectangle with a length of 7 feet and a width of 3 feet.

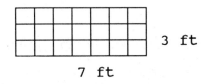

Area of the rectangle = 3 ft x 7 ft = 21 square feet.

CONTINUED NEXT PAGE

EXAMPLE: Calculate the area and perimeter of a garden that is 18' x 24'

AREA = L x W

AREA = 24' x 18' = 432 sq ft

PERIMETER = L + L + W + W

PERIMETER = 24' + 24' + 18' + 18' = 84'

24'

18'

AREA OF A SQUARE = S^2

The area of a square is given by S^2 where S is the measure of a side.
(Actually just Length x Width)

Example: What is the area of a square with a side of 6 feet?

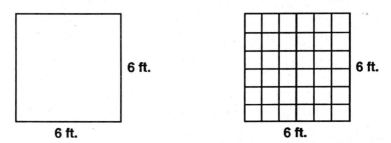

The area is 6' X 6' = 36 square feet which may be written as 36 ft^2.
It is important to use the units and not just the numbers when doing these
~oblems.

AREA OF A TRIANGLE

The *Area of a triangle* is given by the formula: A = 1/2 (B x H), read as: "one-half the base times the height."

This formula doesn't have to be memorized. If you realize that a triangle is half a rectangle, and the area of a rectangle is length times width, just take 1/2 of the area of a rectangle that has a length and width equal to the height and base of the triangle.

Example: Calculate the area of a triangle with a base of 4" and a height of 7".

By formula: A = 1/2 (B x H)

$$A = 1/2 (4" \times 7") = 1/2 (28 \text{ in}^2) = 14 \text{ in}^2$$

The answer is read "fourteen square inches" but *NOT* "fourteen inches squared." (which might produce an answer of 196 sq in.)

JED NEXT PAGE   11

The rectangle and triangle would look like this:

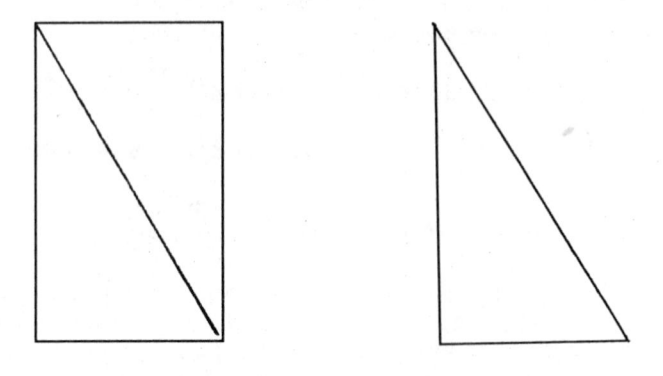

The area of the triangle is one-half that of the rectangle.

AVERAGE (MEAN)   SYMBOL AVE

Definition: The sum of the values of a number of items divided by
              the number of items.

The *average* is a measure of central tendency. Differences from the
average indicate how "far out" a value is. The greater the number of
items, the more representative is the average.

Example: What is the average of the grades 64, 72, 84, 96

            $64 + 72 + 84 + 96 = 316$

            $316/4 = 79$

The average is a statistic used to report rainfall, temperatures, prices,
annual changes. Another measure similar to average is the median
which is the *middle* value. Put the values in order from smallest to
largest and count out to the middle value which is the median.

CIRCLES - EXAMPLES

A circle has a radius of 6 inches. Find its diameter, circumference and area.

$D = 2R$ $\qquad$ $D = 2 \times 6" = 12"$

$C = D \times pi$ $\qquad$ $C = 12" \times 3.14 = 37.68$

$\qquad\qquad\qquad$ $C = 12" \times 22/7 = 37.71$

$A = pi \times R^2$ $\qquad$ $A = 3.14 \times (6")^2 = 113.04$ sq in

$\qquad\qquad\qquad$ $A = 22/7 \times (6")^2 = 113.14$ sq in

The differences in circumference and area calculations is the result of choosing the fractional or decimal equivalents of pi.

CIRCLES - TERMS - AREA

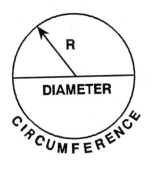

R = RADIUS - the distance from the center of the circle to any point on the circumference

D = DIAMETER - the distance across the circle through the center

C = CIRCUMFERENCE - the distance around the rim of the circle

pi = 3.14 ... ≈ 22/7 (approximations)

pi = the number of diameters that will wrap around one circumference

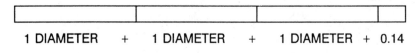

| 1 DIAMETER | + | 1 DIAMETER | + | 1 DIAMETER | + | 0.14 |

$C = D \times pi$ $D = 2R$ $A = pi \times R^2$

17

COMMON DENOMINATOR

Needed to add and subtract fractions with different denominators such as 1/2 + 1/4.

Definition: A number that is evenly divisible by all the denominators in question.

How to get a CD: **Multiply denominators together.** *The product of all the denominators is a common denominator because all the factors HAVE TO divide evenly into the product*

Example: Find a CD for 1/2, 1/3, 1/4, 1/5.

Answer: 2 x 3 x 4 x 5 = 120 is a CD.

Note: 120 is not the LCD (Least Common Denominator) but it IS a common denominator and it will work in all cases.

SEE ALSO: LEAST COMMON DENOMINATOR

COMPARING FRACTIONS

Comparing fractions means to determine which fraction is larger than another. Often the problem requires you to list the fractions *in order* from smallest to largest, or at times, largest to smallest.

EXAMPLE
METHOD 1:

Write in order from smallest to largest:

$$\frac{3}{4} \qquad \frac{2}{3} \qquad \frac{5}{6} \qquad \frac{1}{2}$$

STEP 1:   Find a common denominator - 12

STEP 2:   Write equivalent fractions 9/12, 8/12, 10/12, 6/12

STEP 3:   Write fractions in order by numerator - smallest to largest

$$\frac{6}{12} \qquad \frac{8}{12} \qquad \frac{9}{12} \qquad \frac{10}{12}$$

CONTINUED NEXT PAGE

METHOD TWO:

STEP 1:   Convert all the fractions into equivalent decimals

$3/4 = 0.75$

$2/3 = 0.6666 \ldots$

$5/6 = 0.8333 \ldots$

$1/2 = 0.5$

STEP 2:   Place in order. 0.5, 0.666, 0.75, 0.833

Equivalent to: 1/2, 2/3, 3/4, 5/6

SEE ALSO: EQUIVALENT FRACTIONS, FRACTIONS TO DECIMALS

COMPOSITE NUMBERS

Whole numbers are either *prime* or *composite*. (**Zero or 1 is neither.**)
Composite numbers are products of two or more whole numbers other
than 1 or zero.

Composite numbers are resolved into their prime factors by using the
FACTOR TREE.

EXAMPLE: 24 is "composed" of 2 x 2 x 2 x 3, all of which are prime
numbers.

SEE ALSO: FACTOR TREE, PRIME NUMBERS

CROSS MULTIPLICATION

It is easy to confuse "cross multiplication" and "multiplying across."

CROSS MULTIPLICATION refers to a method of dealing with proportions and equivalent fractions when one of the numbers is missing. The *cross* refers to the X pattern of multiplication. The numerator of one fraction is multiplied by the denominator of the other.

In order to *cross multiply* it is necessary to have an equal sign between two fractions. Cross multiplication is *not* used when multiplying fractions.

EXAMPLE: Find the missing value: $\dfrac{2}{3} = \dfrac{?}{6}$

STEP 1: Multiply 2 x 6 = 12 (cross product)
STEP 2: Divide by 3
Missing value is 4

CONTINUED NEXT PAGE

RULE: *If two fractions are equivalent, their cross products are equal.*

EXAMPLE: Given the equivalent fractions 2/3 = 6/9 note that

2 x 9 = 3 x 6 (2 x 9 is one cross product, 3 x 6 the other)

MULTIPLYING ACROSS **refers to the operation of multiplying fractions.**

RULE: *To multiply two fractions, multiply their numerators; multiply their denominators.*

EXAMPLE: $\dfrac{4}{5}$ x $\dfrac{5}{7}$ = $\dfrac{20}{35}$

SEE ALSO: FRACTIONS-MULTIPLICATION, EQUIVALENT FRACTIONS

DECIMALS - ADDITION AND SUBTRACTION

Rules: LINE UP THE DECIMAL PLACES - tenths with tenths etc.

If a whole number is being added, its decimal point is to the right of the last digit including any zeroes.

Hint: Use graph paper or ruled paper turned sideways as an aid to keep decimal places in line.

Example: Add 5.67 + 3 + 0.007

```
5.67
3           Proper
0.007       setup
8.677
```

If a decimal is subtracted from a whole number, LINE UP THE DECIMAL PLACES. Add as many zeroes to the right of the decimal point as needed, then perform the subtraction.

CONTINUED NEXT PAGE

27

Example: 5 - 0.067

$$
\begin{array}{r}
5.000 \\
-0.067 \\
\hline
4.933
\end{array}
$$

Proper
setup

DECIMALS - DIVISION

Rule: The divisor must be a whole number. To change a decimal divisor to a whole number, the decimal point is "moved." See next page for an explanation of why decimals are "moved."

Example: $10.1 \overline{\smash{)}40.56}$ Step 1. Move decimal of divisor one place to the right.

$101 \overline{\smash{)}405.6}$ Step 2. Move decimal of dividend one place to the right.

$101 \overline{\smash{)}405.6}$ Step 3. Place another decimal above the line.

Step 4. Do the division as you would with whole numbers. Carry out division as many places as requested. If number of places is not requested, carry out three places past the decimal and round off to two.

CONTINUED NEXT PAGE   29

The decimal point is not magically "moved" right or left at whim. In division the reason is quite simple. A division may be expressed as a fraction.

$$10.1\overline{\smash{\big)}40.56}$$ may be expressed as $$\frac{40.56}{10.1}$$

In this fraction problem, the top and bottom are each multiplied by 10 (decimal point moved one place to the right) to form an equivalent fraction with a whole number denominator.

By multiplying by powers of ten, decimals points are "moved."

RESULT: $$101\overline{\smash{\big)}405.60}\frac{4.02}{}$$ (quotient has been rounded)
(to the second place)

SEE ALSO: EQUIVALENT FRACTIONS, ROUNDING DECIMALS, POWERS OF 10

DECIMALS TO FRACTIONS

To change a decimal to a fraction, READ the decimal and WRITE it as you hear it. You need to know the place names of numbers.

Example: 0.24 is read as "twenty-four hundredths." Note the word *hundredths. Hundredths* **is a fraction with a denominator = 100. Therefore 0.24 = 24/100.**

Example: 0.548 is read as five hundred forty-eight thousandths. The denominator is 1000. So, 0.548 = 548/1000.

What about fractions like 1/4? How do they come about from decimals?

Example: 0.25 is twenty-five hundredths. 0.25 = 25/100 = 1/4.

What about mixed numbers?

Example: 1.28 = 1 *and* **28/100**

DECIMALS - MULTIPLICATION

Rules: Do NOT try to line up decimal places. IGNORE the
          decimal points. (It may help to temporarily erase the
          decimal points or rewrite the problem *without* the decimal
          points.)

          Multiply the whole numbers together.

          Count the total number of decimal places in both
          numbers.

          Move decimal in product that many places to left.

Example: 3.47 x 2.14 (4 decimal places)

          347 x 214 = 74258   (count four from right to left)

          Answer: 7.4258

DECIMALS TO PERCENT

To change a decimal to a percent, multiply the decimal by 100%.(Multiply by 100 and add the percent sign to the right of the number.)

Example: **Change 0.16 into a %**

$$0.16 \times 100\% = 16\%$$

100% = 1 because 100/100 = 1

Therefore: 0.16 = 16% = 16/100

The % sign means to divide by 100. What has happened is that 0.16 was multiplied by 100, the % sign was included so that 16% still equals 0.16. There is no change in value, only in form.

Example: **Change 14 into a percent**

$$14 \times 100\% = 1400\%$$

NOTE: **14 = 1400% = 1400/100 = 14**

DECIMALS VS FRACTIONS

There may be differences in final answers between using decimals and fractions when calculating. Some fractions work out evenly into equivalent decimals such as 1/4 = 0.25. However, many fractions *do not* convert evenly such as 1/7 = 0.1428571 ... or 1/3 = 0.33333 (repeating.)

A particular difficulty arises in the use of pi in problems. Pi = 3.14 or 22/7. Both these numbers are *approximations.*

To handle repeating and non-ending decimals, round off the answer to the second place to the right of the decimal unless instructed otherwise. DO NOT round off any decimals before the final answer is found. Errors will creep into the calculations if this is done. *When using a calculator, keep all the places displayed till the very last calculation, then round off.*

Should you use fractions or decimals? Use whatever form feels comfortable, and don't be afraid of converting one into the other.

CONTINUED NEXT PAGE

The decimal system is not symmetrical. There are ONES to the left of the decimal point and TENTHS to the right.

In a WHOLE number, the decimal point is not shown, but really exists to the right of the last digit (including any zeroes.)

The number 3650 can be written 3650. (the decimal point to the right.)

SEE ALSO: FRACTIONS TO DECIMALS

DIVISION FACTS

A division may be expressed as a fraction.

In a fraction, the denominator is the divisor.

Always divide by the number that comes *after* the ÷ sign.

$$\frac{\text{QUOTIENT}}{\text{DIVISOR}/\text{DIVIDEND}}$$

Example: 12 ÷ 14 means $14/\overline{12}$, i.e, 12 divided by 14

Note in this case the divisor is *larger* than the dividend. A common error is to think that one divides the smaller number into the larger. *Any number may be divided by any other number except division by zero.*

NEVER try to divide by zero.

39

EQUIVALENT FRACTIONS

Equivalent fractions are fractions that are equal.

Example: 4/5 = 8/10   (numerator and denominator of
                                  4/5 were multiplied by 2)

Any fraction may be "changed" into *many* equivalent fractions by
multiplying the numerator and denominator by the same quantity.

Example: 1/2 = 2/4 = 3/6 = 4/8 = 5/10 . . .

Another way to think about equivalency is to multiply the fraction by
1 expressed as 2/2 or 3/3 or 4/4 or what have you because any
number divided by itself is 1.   (N/N) except 0/0 which is not allowed.

Uses: Equivalent fractions are useful in finding the "missing"
             value in proportions, simplifying fractions, adding or
             subtracting fractions.

SEE ALSO: PROPORTIONS   41

FACTORS - Prime and Otherwise

When two numbers are multiplied together they form a *PRODUCT*. The numbers that are multiplied together are called *FACTORS*. A given product may have many sets of factors.

$$3 \times 4 = 12 \qquad 3 \text{ and } 4 \text{ are } Factors \text{ of } 12.$$

If a factor cannot be divided evenly by any number other than 1 and itself, the factor is said to be a *PRIME FACTOR*. Prime factors may be found in a Prime Number Table.

The first 10 prime numbers are 2, 3, 5, 7, 11, 13, 17, 19, 23, 29.

Numbers which can be divided evenly by numbers other than 1 or the number itself are called *COMPOSITE*. 12 is composite because it is "composed" of 6 x 2 and can be divided by 6 and 2. Notice that some numbers may be composed of more than one set of factors, but that there is only ONE set of PRIME FACTORS for a given composite number.

SEE ALSO: PRIME NUMBERS, PRIME NUMBER TABLE

FACTOR TREE

A simple and effective way to find all the prime factors of a composite number is to use a *FACTOR TREE.*

1. Find ANY factor that divides into the composite number.

2. Write down the divisor and the quotient. Check to see if either is PRIME. Circle the primes.

3. Continue dividing until *all* the factors are prime.

4. Arrange the circled prime factors in order. The prime factors of 120 are:

2 x 2 x 2 x 3 x 5   OR 2^3 x 3 x 5

SEE ALSO: PRIME NUMBER TABLE

FORWARDS AND BACKWARDS

Going "backwards" is a math skill well worth developing. For example, there are 12 inches in one foot, but it is also true that one foot = 12 inches.

Subtraction and addition are operations that are the reverse of each other. Multiplication and division are also reverse operations.

The concept of a reciprocal is also a "backwards" notion. Multiplying by 1/2 is the same as dividing by 2. 2 = 2/1

Questions like: "what number multiplied by 6 is equal to 30" or "18 is what % of 48" use "backwards" reasoning.

Actually there is nothing backwards about such notions. The reason why they look strange is that math skills are usually taught in one direction only such as 4 x 6 = 24 and not 24 is the product of 6 and what number. The "backwards" skill is worth developing.

SEE ALSO: RECIPROCAL   47

FRACTIONS - ADDITION

Rule: If the denominators are *alike*, just add the numerators and keep the same denominator.

3/5 + 1/5 = 4/5   and   4/7 + 9/7 = 13/7 (OK to leave this way)

Rule: If the denominators are *unlike:*

$$\frac{3}{5} + \frac{1}{5} = \frac{4}{5} \quad / \quad \frac{4}{7} + \frac{9}{7} = \frac{13}{7}$$

-- Find a CD (Common Denominator)
-- Write equivalent fractions
-- Add numerators
-- Keep common denominator

1/2	=	6/12
1/3	=	4/12
1/4	=	3/12

$$\frac{1}{2} + \frac{1}{3} + \frac{1}{4} = \frac{6}{12} + \frac{4}{12} + \frac{3}{12}$$

$$\frac{13}{12}$$

13/12 (OK to leave this way)

FRACTIONS - DEFINITION

A FRACTION IS A DIVISION

The NUMERATOR is always divided by the DENOMINATOR.

TERMS: nUmerator (U for UP.)   Denominator (D for down.)

Always divide by the denominator.

The line between the N and D is a division line.

(D/Down/Denominator/Divide/Divisor)

Whole numbers may be expressed as fractions by dividing by 1

FRACTIONS - DIVISION

YOU DO NOT NEED A CD TO DIVIDE FRACTIONS

Method: -- Express division as a complex fraction
-- Multiply numerator and denominator by reciprocal of the denominator
-- Simplify to lowest terms if possible

Example: $3/4 \div 1/2$

Write as: $\dfrac{3/4 \times 2/1}{1/2 \times 2/1} = \dfrac{6/4}{1} = 3/2$ (a number times its reciprocal = 1)

The rule of thumb "invert and multiply" may be used if understood. The word "invert" does not accurately describe what is happening. The actual word should be *reciprocal*.

$$\frac{3}{4} \cdot \frac{2}{1} = \frac{6}{4} = \frac{3}{2}$$

SEE ALSO: RECIPROCALS

FRACTIONS - MULTIPLICATION

YOU DO NOT NEED A CD FOR MULTIPLICATION

Rule: **Multiply numerators, multiply denominators. It does not matter how many fractions there are, just multiply all the numerators together, and all the denominators together. As a last step, simplify to lowest terms.**

$$3/5 \times 4/9 = 12/45 = 4/15$$

Multiplication of fractions can be understood better if the X of multiplication is changed to the word "of". What is 1/2 of 1/2? The answer 1/4, is easier to see than "timesing" 1/2 by 1/2.

At times it is possible to simplify a multiplication. Use the first rule of fractions: *You may divide the numerator and denominator of any fraction by the same quantity without changing the value of the fraction.*

CONTINUED NEXT PAGE

5/8 x 8/9 x 9/15 = 1/3

$$\frac{\cancel{5}^{1}}{\cancel{8}_{1}} \cdot \frac{\cancel{8}^{1}}{\cancel{9}} \cdot \frac{\cancel{9}^{1}}{\cancel{15}_{3}} = \frac{1}{3}$$

The top and bottom is divided by 5, then 8, then 9. There is nothing left to simplify.

$$\frac{\cancel{5}^{1}}{\cancel{8}_{1}} \times \frac{\cancel{8}^{1}}{\cancel{9}_{1}} \times \frac{\cancel{9}^{1}}{\cancel{15}_{3}} = \frac{1}{3}$$

In CHAIN-MULTIPLICATIONS remember to multiply by numerators and divide by denominators

FRACTIONS - SUBTRACTION

Rule: Fractions with the same denominator -- subtract numerators.
      Keep the same denominator.

Example: 5/6 - 2/6 = 3/6   Simplified: 1/2

$$\frac{5}{6} - \frac{2}{6} = \frac{3}{6} = \frac{1}{2}$$

Rule: Fractions with different denominators

Step 1. Find a common denominator (CD)
Step 2. Write equivalent fractions with a CD
Step 3. Subtract numerators
Step 4. Write difference over the CD

EXAMPLE: $\dfrac{2}{3}$ $\dfrac{4}{9}$ CD = 9

$\dfrac{6}{9}$ $\dfrac{4}{9}$ = $\dfrac{2}{9}$

FRACTIONS TO DECIMALS

To change a fraction into a decimal - divide the numerator by the denominator.

There are two ways the result can appear.

-- the decimal will come out even and stop
-- the decimal will repeat

Examples: 1/8 = 0.125 (terminal)
              1/3 = 0.33333333 (repeating)
              1/7 = 0.142857142857 (repeating)

With the repeating decimals, round off to the second place to the right of the decimal point.

The common fractions have common decimals that should become familiar to the student such as 1/4 = 0.25; 1/2 = 0.5, 1/5 = 0.2

SEE ALSO: DECIMALS TO FRACTIONS

FRACTIONS TO PERCENTS

Rule: To change a fraction to a percent, first change the fraction to a decimal, then change the decimal to a percent.

Example: Change 3/5 into a percent.

   3/5 = 0.6

   0.6 x 100 = 60

   Add the % sign = 60%

Example: Change 1 into a percent

   1/1 = 1; 1 x 100 = 100; add the % sign = 100%

Examples: 1/4 = 0.25 = 25%
            1/10 = 0.10 = 10%
            3/1 = 3.0 = 300%

SEE ALSO: FRACTIONS TO DECIMALS - TO PERCENT   61

GREATEST COMMON FACTOR

The Greatest Common Factor (GCF) is a the largest whole number that will divide evenly into each of the numbers in a set. There are two conditions here. First, the GCF is <u>common</u> to all the numbers in question, and second it is the <u>largest</u> divisor possible.

EXAMPLE: Find the GCF of 15, 30 and 60.

By looking at the numbers it is easy to see that 15 is the *largest* factor that will divide into all three numbers.

In problems where the GCF is not evident, the following method may be used:

Rules: Find all the prime factors of each number.
Express these factors in exponential form.
Multiply together the *lowest* exponentials of each *common* base.
The result is the GCF.

CONTINUED NEXT PAGE

EXAMPLE: Use the method described above to find the GCF of 15, 30 and 60

1. Prime factors of 15 are 3 x 5
   Prime factors of 30 are 2 x 3 x 5
   Prime factors of 60 are 2 x 2 x 3 x 5 or 2^2 x 3 x 5

2. Lowest *common* exponentials are 3^1, 5^1

3. 3 x 5 = 15 which is the GCF

EXAMPLE: Use the same method to find the GCF of 40, 50, 60, 120

1. Prime factors of 40 are 2 x 2 x 2 x 5 or 2^3 x 5^1
   Prime factors of 50 are 2 x 5 x 5 or 2^1 x 5^2
   Prime factors of 60 are 2 x 2 x 3 x 5 or 2^2 x 3^1 x 5^1
   Prime factors of 120 are 2 x 2 x 2 x 3 x 5 or 2^3 x 3^1 x 5^1

2. Lowest common exponentials are 2^1, 5^1

3. 2 x 5 = 10 which is the GCF

LEAST COMMON DENOMINATOR

The Least Common Denominator is useful, but not essential for adding or subtracting fractions. The COMMON DENOMINATOR works splendidly. The LCD is just a sophistication of arithmetic.

Definition: The *LCD is the smallest* whole number that is evenly divisible by all the denominators in the problem.

How to find: Method 1 - INSPECTION. Look at the numbers and pick out the LCD. "Figure it out."

Example: Find the LCD for 1/24, 1/12, 1/6, 1/3

By "inspection" (look at the numbers) it can be seen that 24 is the smallest number divisible by 24, 12, 6 and 3.

Some LCDs are best found by inspection. Even denominators and multiples of numbers are key items to spot.

Method 2 - FACTOR TREE METHOD. (next page)   65

-- Use the factor tree to find the prime factors of each of the denominators.

-- Multiply together the highest power of each different base.

Example: Find the LCD for 1/24, 1/12, 1/6, 1/3

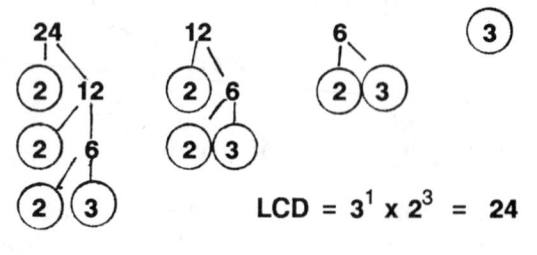

$$LCD = 3^1 \times 2^3 = 24$$

Do not use all the different powers of 2, only the biggest. The smaller powers are duplicates and if used, will make the LCD an ordinary CD.

SEE ALSO: COMMON DENOMINATOR

LEAST COMMON MULTIPLE

The Least Common Multiple (LCM) of a set of whole numbers is the smallest multiple that is common to the multiples of each number.

EXAMPLE: What is the LCM of 5 and 6?

Multiples of 5 are 5, 10, 15, 20, 25, 30 . . .
Multiples of 6 are 6, 12, 18, 24, 30 . . .

The lowest multiple that is common to both sets is 30.

Finding the answer by inspection will not always be as easy as in the above example. A method of determining the LCM follows:

1. Write the prime factors of each number in exponential form.
2. Multiply together the *highest* exponential of each *different* base.
3. The result is the LCM.

CONTINUED NEXT PAGE

EXAMPLE: Find the LCM of 105 and 210.

1. The prime factors of 105 are 3 x 5 x 7
   The prime factors of 210 are 2 x 3 x 5 x 7
2. Highest exponentials of each different prime are
   2 x 3 x 5 x 7
3. The result is 210

EXAMPLE: Find the LCM of 24, 36 and 48

1. Prime factors of 24 are 2^3 x 3
   Prime factors of 36 are 2^2 x 3^2
   Prime factors of 48 are 2^4 x 3
2. Highest exponentials of each different prime are

   2^4 x 3^2
3. The result is 144

SEE ALSO: GREATEST COMMON FACTOR

METRIC SYSTEM

The metric system is based on the METER which is 39.37 inches. The distance from the equator to either pole is 10 million meters. Liquid measure is based on the LITER, approximately 1.0567 quarts. Latin prefixes describe parts or multiples of a meter and liter in terms of powers of ten.

1000 meters is a KILOmeter	1000 liters is a KILOliter
100 meters is a HECTOmeter	100 liters is a HECTOliter
10 meters is a DECAmeter	10 liters is a DECAliter
1 meter is 39.37 inches	1 liter is 1.0567 quarts
0.1 meter is a DECImeter	0.1 liter is a DECIliter
0.01 meter is a CENTImeter	0.01 liter is a CENTIliter
0.001 meter is MILLImeter	0.001 liter is a MILLIliter

It is a convenient fact that I milliliter (ml) = 1 cubic centimeter (cc) which relates liquid measure and volume.

Conversion factors that are handy are:

2.54 cm = 1 inch   1.06 qts = 1 liter   2.2 lbs = 1 kg   **69**

MIXED NUMBERS - ADDITION

Mixed numbers are simply numbers composed of a whole number part and a fraction part.

Example: 4 1/2 is four PLUS one-half

It is NOT necessary to change mixed numbers into improper fractions. Treat the whole numbers separately from the fractions and then combine.

Example:

Add: 5 1/3 and 2 1/5

First add the 5 and 2 to get 7 (whole number part)

Then add 1/3 and 1/5 to get 8/15   (need a CD)

Then put them together 7 8/15

If the fractional part of the answer is an improper fraction, change the improper fraction to another mixed number and add again.

CONTINUED NEXT PAGE

71

EXAMPLE: Add   1 4/5 + 6 2/3

First add the 1 and 6 to get 7 (whole number part)

Then add 4/5 + 2/3 to get 22/15 (need a CD)

Note that 22/15 is an improper fraction

Change 22/15 into a mixed number 1 7/15

Add 1 7/15 + 7 to get 8 7/15 (final answer)

SEE ALSO: IMPROPER FRACTIONS, COMMON DENOMINATOR

MIXED NUMBERS - DIVISION

Steps: 1. Change mixed numbers into improper fractions.

       2. Divide the first fraction by the second.

       3. Simplify if possible.

Example: $4\ 1/8 \div 2\ 1/5$

        $4\ 1/8 = 33/8;\quad 2\ 1/5 = 11/5$

        $33/8 \div 11/5 = 33/8 \times 5/11 = 15/8$

Simplify: $15/8 = 1\ 7/8$

SEE ALSO: FRACTIONS - DIVISION   73

MIXED NUMBERS - MULTIPLICATION

Rules:
1. Change mixed numbers into improper fractions.

2. Multiply fractions by multiplying numerators together and denominators together.

3. Simplify if necessary.

Example: 4 1/2   x   3 1/3

9/2   x   10/3   =   90/6

Simplify: 30/2   =   15/1   =   15

SEE ALSO: FRACTIONS - MULTIPLICATION   75

MIXED NUMBERS TO IMPROPER FRACTIONS

Definition: An *improper fraction* is one in which the numerator is greater than or equal to the denominator. There is nothing wrong with an improper fraction. These fractions behave just the same way as proper fractions.

Steps: -- write the whole number part as an equivalent fraction with the same denominator as the fraction part
              -- add the two fractions together

Example: Change 2 1/2 to an improper fraction.

-- the 2 is 2/1. It may be expressed as 4/2 (CD with the 1/2 part)
-- 4/2 and 1/2 = 5/2

The "magic formula" of multiplying the whole number by the denominator of the fraction and then adding the numerator is exactly what is happening here. See next page.

77

EXAMPLE: Change 2 1/3 into an improper fraction

STEP 1: Multiply the whole number part times the
         denominator 2 x 3 = 6

STEP 2: Add the numerator   6 + 1 = 7

STEP 3: Write the new sum over the denominator to
         get 7/3

Why is this method valid?

First, a mixed number is an addition of a whole number and a fraction.
A whole number may be expressed as a fraction by writing it over 1.
2/1 can be changed into the equivalent fraction 6/3. Now add the 6/3
and the 1/3 to get 7/3.

When the whole number 2 is multiplied by the denominator 3, you are
expressing how many thirds there are in 2. There are 6 thirds in two.
Now add the 1 of the 1/3 to get seven thirds.

SEE ALSO: EQUIVALENT FRACTIONS

MIXED NUMBERS - SUBTRACTION

There are several ways to approach the subtraction of mixed numbers. The first method is not the easiest to do, but *is* the easiest to remember. The remaining case deserves attention because it affords a deeper understanding of mixed numbers.

TASK: Subtract: 4 1/2 - 2 5/6

METHOD 1: - Change both mixed numbers into improper fractions.
           - Find a common denominator (CD)
           - Subtract second numerator from first numerator

EXAMPLE: 4 1/2 = 9/2;   2 5/6 = 17/6   improper fractions

           9/2 = 27/6;   17/6 = 17/6   both have a CD

           27/6 - 17/6 = 10/6   subtract numerators

           10/6 = 1 4/6 = 1 2/3   (simplified)

CONTINUED NEXT PAGE   79

METHOD 2: Line up subtraction vertically.

    4 1/2   Note that 5/6 is larger than 1/2.
    -2 5/6   You must do *three* things. First change
             1/2 into 3/6 (Equivalent fraction) next
             change the 4 into 3 and 6/6. Third, add
             6/6 to 3/6 to get 9/6. Now the top fraction
is bigger than the bottom fraction and the subtraction may proceed.

    3 9/6   (Equivalent to 4 1/2 but with a CD)
    -2 5/6   (bottom fraction is OK - don't change it)
    1 4/6   = 1 2/3 simplified

SPECIAL CASE: Mixed number subtracted from a whole number.

EXAMPLE: 4 - 2 1/3   Change the 4 into 3 and 3/3 then
                             subtract as above.

    3 3/3
    -2 1/3
    1 2/3

SEE ALSO: FRACTIONS - SUBTRACTION, CD, SIMPLIFYING

MULTIPLICATION TABLE

1	2	3	4	5	6	7	8	9	10	11	12	13	14	15
2	4	6	8	10	12	14	16	18	20	22	24	26	28	30
3	6	9	12	15	18	21	24	27	30	33	36	39	42	45
4	8	12	16	20	24	28	32	36	40	44	48	52	56	60
5	10	15	20	25	30	35	40	45	50	55	60	65	70	75
6	12	18	24	30	36	42	48	54	60	66	72	78	84	90
7	14	21	28	35	42	49	56	63	70	77	84	91	98	105
8	16	24	32	40	48	56	64	72	80	88	96	104	112	120
9	18	27	36	45	54	63	72	81	90	99	108	117	126	135
10	20	30	40	50	60	70	80	90	100	110	120	130	140	150
11	22	33	44	55	66	77	88	99	110	121	132	143	154	165
12	24	36	48	60	72	84	96	108	120	132	144	156	168	180
13	26	39	52	65	78	91	104	117	130	143	156	169	182	195
14	28	42	56	70	84	98	112	126	140	154	168	182	196	210
15	30	45	60	75	90	105	120	135	150	165	180	195	210	225

(Note diagonal of perfect squares)

81

NUMBERS - KINDS OF

COUNTING - 1, 2, 3, 4 ...

DECIMAL - numbers expressed in powers of 10, such as 5.37

FRACTIONS - the quotient of a/b where b is *not* zero

IRRATIONAL - any number that cannot be expressed as a fraction such as 5.34237 ... (goes on and on)

MIXED NUMBERS - combination of a whole number and a fraction

NATURAL NUMBERS - another name for the counting numbers

NEGATIVE NUMBERS - any number less than zero, found to the left of zero on the number line

RATIONAL - any number that can be written as a fraction

WHOLE NUMBERS - 0, 1, 2, 3, 4 ...

83

OPERATIONS AND OPERATORS

The *BASIC OPERATIONS* of arithmetic are adding, subtracting, multiplying, and dividing. Other operations are squaring, cubing, taking a square root, a cube root and percentage.

The *OPERATORS* of arithmetic are the symbols $+$, $-$, X, $\div$, the radical sign ($\sqrt{}$), %, and exponents.

The *order* in which the operations are done is critical.

The operators *operate* on numbers. They permit numbers to interact with one another according to the rules of addition, subtraction, multiplication, division, etc.

SEE ALSO: ORDER OF OPERATIONS, OPERATIONS - TERMS 85

OPERATIONS - TERMINOLOGY

Addition - "answer" is called the SUM

Cubing - "answer" is called the CUBE

Division - "answer" is called the QUOTIENT

Multiplication - "answer" is called the PRODUCT

Square root - "answer" is called the SQUARE ROOT

Squaring - "answer" is called the SQUARE

Subtraction - "answer" is called the DIFFERENCE

SEE ALSO: OPERATIONS AND OPERATORS, SQUARE ROOT   87

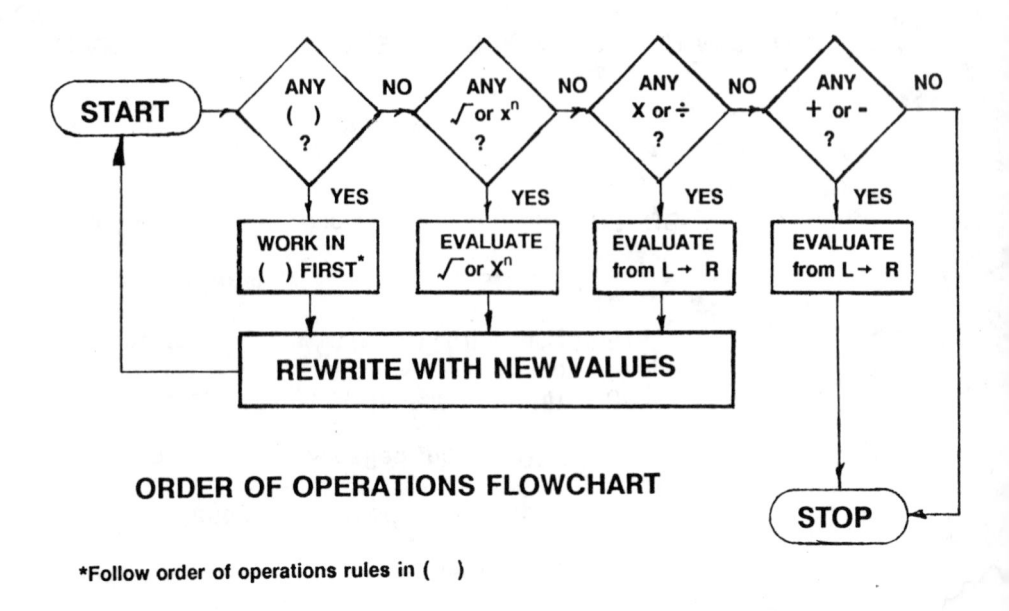

ORDER OF OPERATIONS FLOWCHART

*Follow order of operations rules in ( )

ORDER OF OPERATIONS

The *order* in which operations are performed in arithmetic is crucial.
If the rules of order are not followed, calculations will come out wrong.
The *Order of Operations* is:

First:	**Do everything that is in parentheses** *
Second:	**Evaluate numbers with exponents and radicals**
Third:	**Do all the multiplications and/or divisions (whichever come first) from left to right.**
Fourth:	**Do all additions and/or subtractions (whichever come first) from left to right.**

Example: $16 - 12 \div 2^2 \times 3 = 7$

Step 1: Check for ( ) - No parentheses, go to step 2
Step 2: Evaluate the exponential 2 squared which is 4
Step 3: Divide 12 by 4 to get 3
Step 4: Multiply 3 by 3 to get 9
Step 5: Subtract 9 from 16 to get 7

* Follow the Order of Operations in the parentheses also.

PERCENT   Symbol   %

The percent sign (%) means to *divide by 100*. On a calculator, the %
button divides the number in the display by 100. The % sign can be
treated as an operator -- anywhere you see the % sign you can divide
the preceding number by 100 and drop the % sign.

EXAMPLE: 40% means 40 $\div$ 100 = 40/100 = 0.4

40% of 50 means

40/100 x 50 = 2000/100 = 20 OR

0.4 x 50 = 20

To "take the percentage" of a number, just multiply the two numbers
together and divide by 100.

Thinking of % as a division by 100 will clarify a great deal of
unnecessary mystery.

91

PERCENT CHANGE

DEFINITION: $\dfrac{\text{The amount of change}}{\text{original}} \times 100 = \%$ change

STEP 1. Determine the amount of change (usually the difference between the two numbers given)

STEP 2. Determine the original amount

STEP 3. Divide the amount of change by the original

STEP 4. Multiply the quotient in STEP 3 by 100 to get the % change

EXAMPLE: The temperature changed from 33° to 45° in one hour. What was the percent increase in temperature? The change is 12°. The original is 33°. The % change is

$12/33 \times 100 = 36.4\%$ increase (rounded to nearest tenth)

PERCENT FRACTION

The *percent fraction* is simply a fraction that has a denominator of 100. By looking at the numerator you can determine the percentage equivalent of the fraction.

EXAMPLE: $60/100 = 60\%$ because the % sign means to /100, i.e., divide by 100.

EXAMPLE: $4/100 = 4\%$

EXAMPLE: $500/100 = 500\%$

EXAMPLE: $0.1/100 =$ one-tenth of a percent

If a fraction can be converted into an equivalent fraction with a denominator of 100, the percent value can be read directly from the fraction.

EXAMPLE: $1/2 = 5/10 = 50/100 = 50\%$

SEE ALSO: PERCENT, FRACTION TO PERCENT

95

PERCENTS TO DECIMALS

To change a percent to a decimal, divide the number by 100 and remove the % sign.

What is happening: The % sign means to divide by 100, so use it as an operational sign. When the operation is done (the division by 100) the % sign is "used up," just like the + sign "disappears" as in the problem 2 + 3 = 5. The plus sign gets "used up."

Examples: 40% = 40/100 = 0.4

3% = 3/100 = 0.03

300% = 300/100 = 3

Note: All three forms, %, fraction and decimal are EQUIVALENT and may be used interchangeably.

SEE ALSO: PERCENT, PERCENTS TO FRACTIONS, DECIMALS TO PERCENTS

PLACE NAMES OF NUMBERS

2, 4 1 7, 8 7 0 . 4 2 7, 5 9 6

Millions	Hundreds of thousands	Tens of thousands	Thousands	Hundreds	Tens	Units	Tenths	Hundredths	Thousandths	Ten-thousandths	Hundred-thousandths	Millionths
2	4	1	7	8	7	0	4	2	7	5	9	6

NOTICE:
  -- Number places less than one end in *THS* such as ten*ths.*
  -- The decimal point is read as "and"
  -- There are hyphens in the ten-thousandths and hundred-thousandths places to distinguish them from ten thousandths and a hundred thousandths

EXAMPLE: 0.010 is ten thousandths while 0.0001 is one ten-thousandth

POWERS OF NUMBERS

Numbers may be "raised to *powers.*" The little number to the upper right of a number is called its *POWER* (exponent.) Example:

2^3 is read as " Two raised to the third power."

Two is called the *BASE*, the 3 is the *POWER.* 2^3 means that the base 2 is used as a factor (multiplier) 3 times[*] written as:

$2 \times 2 \times 2 = 8$

So we say "two to the third power is eight."

SPECIAL NAMES: A number raised to the 2nd power is *SQUARED.* A number raised to the 3rd power is *CUBED.*

[*] To say "two times itself three times" is incorrect. The result would be 16.

ANY NUMBER MAY BE RAISED TO ANY POWER[*]

Whole numbers: $4^3 = 64$; $0^3 = 0$

Decimals: $1.3^2 = 1.69$; $0.02^3 = 0.000008$

Negative numbers: $(-3)^2 = 9$; $(-2)^3 = -8$

Fractions: $(1/2)^2 = 1/4$; $(1/3)^3 = 1/27$

Mixed numbers: $(1\text{-}1/2)^2 = 9/4$; $(2\text{-}1/3)^2 = 49/9$

The ZERO power: $4^0 = 1$; $11^0 = 1$

The FIRST power: $3^1 = 3$; $5^1 = 5$

Negative powers: $2^{-3} = 1/8$; $3^{-2} = 1/9$

Decimal powers: $3^{1.2} = 3.74$; $4^{2.4} = 27.86$ **(done on a calculator)**

[*] **Exception:** 0^0 **can't be done**

POWERS OF TEN

Symbol 10^n

Many practical problems involve multiplying or dividing by powers of ten. Metric measurement, order of magnitude, currency calculations are examples.

RULE: When multiplying by powers of 10, move the decimal point one place to the right for each factor of ten. Notice that the resulting number gets larger.

EXAMPLE: $60 \times 10^3 = 60 \times 10 \times 10 \times 10 = 60,000$

EXAMPLE: $0.5 \times 10^2 = 0.5 \times 10 \times 10 = 50$

EXAMPLE: $6.153 \times 10^3 = 6.153 \times 10 \times 10 \times 10 = 6153$

The decimal was moved one place to the right for each factor of 10.

CONTINUED NEXT PAGE

103

RULE: When dividing by powers of 10, move the decimal point one place to the left for each factor of ten. Notice that the resulting number gets smaller.

EXAMPLE: $6,000 \div 10^3 = 6$

EXAMPLE: $138.2 \div 10^4 = 0.01382$

WHY do you "move decimals?" Answer: To multiply or divide by powers of ten.

WHEN do you "move decimals?" Answer: When dividing by a decimal, when changing a decimal to a %, when changing a % to a decimal.

WHAT does "moving a decimal point" do? Answer: moving it one place to the right makes the number ten times larger. Moving it one place to the left makes the number ten times smaller.

SEE ALSO: DECIMALS - DIVISION, DECIMAL TO PERCENT, PERCENT TO DECIMAL

PRIME NUMBERS - DEFINITION

DEFINITION: A prime number is a whole number that can be evenly divided only by itself and 1.

EXAMPLES: 5 is prime
7 is prime
10 is *not* prime (it is a multiple of 5)

The number 2 is the only *even* prime number, all other even numbers are *composite*, i.e., the product of other numbers. A *prime* number can only be expressed as the product of itself and 1.

The number 1 is not prime by definition. The number 0 is not prime.

SEE ALSO: PRIME NUMBER TABLE, COMPOSITE NUMBERS

PRIME NUMBERS TABLE - Between 1 and 1000

2	3	5	7	11	13	17	19	23	29	31	37
41	43	47	53	59	61	67	71	73	79	83	89
97	101	103	107	109	113	127	131	137	139	149	151
157	163	167	173	179	181	191	193	197	199	211	223
227	229	233	239	241	251	257	263	269	271	277	281
283	293	307	311	313	317	331	337	347	349	353	359
367	373	379	383	389	387	401	409	419	421	431	433
439	443	449	457	461	463	467	479	487	491	499	503
509	521	523	541	547	557	563	569	571	577	587	593
599	601	607	613	617	619	631	641	643	647	653	659
651	673	677	683	691	701	709	719	727	733	739	743
751	757	761	769	773	787	797	809	811	821	823	827
829	839	853	857	859	863	877	881	883	887	907	911
919	929	937	941	947	953	967	971	977	983	991	997

PROPORTIONS

A proportion is a comparison between two equivalent ratios.

   1/4 = 6/24 is read as "one fourth equals six twenty-fourths."

A proportion is a set of two equivalent fractions.

A proportion has an = sign between two fractions.

Rule: The cross-products of a proportion are equal.

Example: 1 x 24 = 6 x 4

Proportion problems usually involve finding one of the missing parts.

There are two major methods of finding the value of the ?.

Example: 4/7 = ?/14

CONTINUED NEXT PAGE

Method 1: Equivalent fractions

Note that by multiplying the denominator 7 by 2 produces the product (14) of the second ratio. So, multiply the 4 x 2 to get the missing numerator.

Method 2: Cross-multiplication

$$\frac{4}{7} = \frac{?}{14} \qquad \frac{14 \times 4}{7} = 8$$

Note that an = sign is necessary to do a cross-multiplication. It is wrong to try to cross-multiply when multiplying fractions together.

When proportions include units such as inches and feet, carry the units along with the numbers.

PYTHAGOREAN THEOREM

Rule: *The sum of the squares of the sides of a right triangle is equal to the square of its hypotenuse.* **Expressed by the formula:**

$$a^2 + b^2 = c^2$$

where a and b are sides and c is the hypotenuse.

Uses: **Relates parts of a right triangle so that if two parts are known, the third can be calculated. Used in navigation, surveying, engineering, construction.**

Example: **One side is 3 feet, the other side is 4 feet, what is the length of the hypotenuse?**

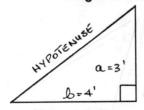

$$3^2 + 4^2 = h^2$$

$$9 + 16 = 25$$

$$h = 5$$

111

RADICALS

SYMBOL $\sqrt{\ \ }$

The radical sign $\left(\sqrt{\ \ }\right)$ represents the operation of extracting a *root* of a number.

Although there are many roots of a number, the common roots are square and cube roots.

The radical sign is composed of two parts, the bracket itself $\sqrt{\ }$ and the *index*, a small number sometimes found in the little "v" of the radical sign. If there is no index shown, it is assumed to be 2 which means you are looking for the second root commonly called the square root.

Example: $\sqrt{36} = 6$ "the square root of 36 is 6"

Example: $\sqrt[3]{8} = 2$ "the cube root of 8 is 2"

SEE ALSO: SQUARE, SQUARE ROOTS, SQUARE ROOT TABLE **113**

RATIO

A RATIO may be expressed as a fraction. The fraction may be proper or improper. The rules of fractions apply to ratios.

Ratios are COMPARISONS.

Example: There are 4 men and 5 women in a group.

The ratio of MEN to WOMEN is: 4 men/5 women

The ratio of WOMEN to MEN is: 5 women/4 men

The ratio of MEN to TOTAL is: 4 men/9 in group

Ratios may be COMPLEX fractions such as 1/8 ounce to 4/5 lbs.

$$\frac{1/8 \text{ oz}}{4/5 \text{ lb}} = 1/8 \times 5/4 \qquad \frac{\text{oz}}{\text{lb}} = \frac{5 \text{ oz}}{32 \text{ lb}}$$

RECIPROCALS

It is easier to demonstrate reciprocals than to define them.

2/3 is the reciprocal of 3/2

3/2 is the reciprocal of 2/3

2 is the reciprocal of 1/2 (2 = 2/1)

Reciprocals occur in *pairs*. One number is the *reciprocal* of the other.

Definition: A number multiplied by its reciprocal = 1.

2/3 x 3/2 = 1; 2/1 x 1/2 = 1

Reciprocals are used for clearing fractions, dividing fractions by fractions, in physics and in technical mathematics such as electronics calculations.

Thought: Multiplying by 1/2 is the same as dividing by 2.   117

ROUNDING DECIMALS

Task: Round 4.106 to the nearest "tenth"

Skills: Must know place names such as tenths, hundredths, thousandths

Rules:
1. Locate the place to be rounded

2. Look to the neighbor to its right

3. If the neighbor is 5 or greater, raise the value of the place to be rounded one number higher

4. If the neighbor is less than 5, drop all the digits following the place to be rounded

Example above: The tenth place is 1, the neighbor to its right is 0 so drop the digits 0 and 6. Rounded = 4.1

CONTINUED NEXT PAGE

EXAMPLE 1:   Round 4.875 to the nearest *tenth*

The *tenth* place is the digit 8

The neighbor to the right of 8 is a 7

7 is larger than 5

Round UP to the higher digit   ANS: 4.9

EXAMPLE 2:   Round 4.875 to the nearest *hundredth*

The *hundredth* place is a 7

The neighbor to the right of 7 is a 5

Round UP to the higher digit ANS: 4.88

SEE ALSO: ROUNDING WHOLE NUMBERS

ROUNDING WHOLE NUMBERS

Task: "Round 2546 to the nearest hundred."

Skills needed: Must know *names* of number places such as "tens, hundreds." Must be able to *order* numbers.

Find the nearest hundred *smaller* than 2546 and the nearest hundred *larger* than 2546.

> **2500** nearest *smaller* hundred
> *2546*
> **2600** nearest *larger* hundred

Now, by looking at the numbers, determine whether 2546 is closer to 2500 or to 2600. The number being tested is closer to 2500, therefore we say 2546, rounded to the nearest hundred is 2500.

Suppose the number falls right in the *middle*, such as 2550? Rule: If the number falls exactly in the middle, *go higher.*

121

CONTINUED NEXT PAGE

ROUNDING RULE OF THUMB

Rule: To round a whole number, go to the place to be rounded and inspect the digit to the right. If that digit is 5 or greater, round up (to the higher number.) If that digit is less than 5, change the remaining digits to the right to zeroes.

Example: Round 2546 to the nearest hundred.

Step 1.   Go to the hundreds place. It is the 5.

Step 2.   Inspect the digit to the right. It is 4.

Step 3.   Since 4 is smaller than 5, round down (to lower number.)

Step 4.   Change the 4 and 6 to zeroes.

2546 rounded to the nearest hundred is 2500.

SCIENTIFIC NOTATION

Scientific Notation is a method of expressing very large and very small numbers in terms of powers of 10. For example, there are 602,000,000,000,000,000,000,000 molecules of H_2O in 18 grams of water (Avagadro's Number.) A more convenient way to write the number is 6.02×10^{23}.

The rules for writing in scientific notation are:

1. Locate the decimal point in the original number.
2. Count the number of places till you arrive to the right of the first significant figure (not including zeroes.)
3. This count of places is the exponent of 10.
4. If you counted from right to left, (large number) the exponent is positive.
   If you counted from left to right, (small number) the exponent is negative.
5. Round off the remainder of the number to three significant figures.

EXAMPLE: $0.000,000,00237 = 2.37 \times 10^{-9}$

SIMPLIFYING (fractions, mixed numbers, etc.)

To "simplify" in mathematics means many things.

FRACTIONS: Simplify means to express the fraction in lowest terms, i.e., with a numerator and denominator that are not divisible by any number other than 1. You must find a number that divides evenly into both the numerator and the denominator and continue doing this till no further simplification is possible. Use a factor tree. SEE: FACTOR TREE

IMPROPER FRACTIONS: Here, simplify means to change the improper fraction into a mixed number by dividing the denominator into the numerator and expressing the remainder as a fraction. Example: 22/5 = 4 2/5.

NUMERICAL EXPRESSIONS: Perform as many of the indicated operations as possible following the order of operations. When no more operations may be performed, the expression is said to be simplified.

SEE ALSO: MIXED NUMBERS TO IMPROPER FRACTIONS

SQUARES

A square can be a geometrical figure that has all four sides equal and all the angles right angles.

A square can also be a number that has been multiplied by itself. Example: 4 x 4 = 16. 16 is said to be the "square" of 4. There is an important connection between these two definitions.

If numbers are represented as boxes with the width as one factor and the length as the other, the box will come out either a square or a rectangle. Example: 3 x 11 = 33 is a rectangle, while 3 x 3 is a square. Those whole numbers that can be represented as squares are called "perfect squares."

SEE ALSO: SQUARE AND RECTANGULAR NUMBERS

SQUARES AND SQUARE ROOTS - Uses

A number is squared to find the area of a square when given a side. Many physical laws in nature obey the *square rule* such as the way the intensity of light varies with distance. Distances are often calculated by squaring the sides of a triangle. In construction, squaring is used to make sure that angles in buildings are "square" such as foundations, walls, pilings.

Square roots are used to find the side of a square if you know the area. One of the most important uses is in finding the parts of a right angle triangle using the Pythagorean Theorem. The ancient mathematicians knew that if you square the hypotenuse (longest part) of a right angle triangle, it would equal the sum of the squares of the other two sides.

Example: Given the triangle with sides 3 and 4 and a hypotenuse of 5,

$$a^2 + b^2 = c^2$$

$3^2 = 9 \quad 4^2 = 16$

$9 + 16 = 25$

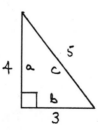

Twenty-five is the sum of the squares of 3 and 4

SQUARE AND RECTANGULAR NUMBERS

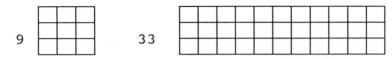

9   9 is a square number   33   33 is a rectangular number

To "square" a number means to multiply the number by itself. The *AREA* of a square is *"Side Squared."* The multiplication table contains a diagonal of perfect squares. The *AREA* of a rectangle is *"length times width."*

Decimals can also be squared. $3.1 \times 3.1 = 9.61$
Be careful with the decimal place.

SEE ALSO: TABLE OF SQUARES AND ROOTS   131

SQUARE ROOTS - Definitions   SYMBOL $\sqrt{}$

-- The square root of a number may be thought of as the side of a square that has an area equal to the number. Example: A square with an area of 25 square units has a side of 5 units.

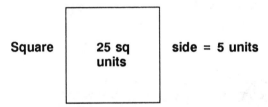

Square   25 sq units   side = 5 units

-- The square root of a number may be thought of as the reverse of squaring. The number 4 squared is 16. The square root of 16 is 4. Q. What number multiplied by itself is 49? A. 7. Therefore, seven is the square root of 49.

CONTINUED NEXT PAGE

The "Common Square Roots" are the roots of the "perfect squares." They also happen to be the counting numbers:

$$\sqrt{1} = 1 \qquad \sqrt{16} = 4$$

$$\sqrt{4} = 2 \qquad \sqrt{25} = 5$$

$$\sqrt{9} = 3 \qquad \sqrt{36} = 6$$

Note: A square root in NOT a division. Look closely at the difference between a square root sign (called a radical sign) and the division symbol.

$$\sqrt{} \qquad\qquad \diagup\overline{}$$

Square root   Division

SQUARE ROOTS - Helpful hints

Most square roots do not work out evenly. $\sqrt{5}$ = 2.236... it keeps going and going and going. How far you carry it out depends on the precision of your calculation. For normal use, round off two places right of the decimal. $\sqrt{5}$ = 2.24

Some square roots look difficult, but are really simple squares in disguise. Example: What is the square root of 1.44? Notice that without the decimal point, the number would be 144, a perfect square. So try 1.2 x 1.2. It works out. However, the square root of 14.4 does not come out evenly. Try it on a calculator.

The diagonal of perfect squares in the multiplication table is also a square root table of the perfect squares. Locate a number on the diagonal and its square root will be found at the top of the column and at the left hand end of the row.

SQUARE ROOTS - How to find them

A square root may be found in a number of ways:

1. Look it up in a *Table of Square Roots.*

2. Use a hand calculator (enter number, then depress the square root key.)

3. Approximate by guessing, then square the number to check out the approximation. Repeat until the product is as close to the square as desired.

4. Raise the number to the 0.5 power on a calculator.

5. Of fractions: find square root of the numerator, find square root of the denominator.

SQUARE ROOTS - Table of Squares and Roots

NO	SQ	SQ RT	NO	SQ	SQ RT
1	1	1	17	289	4.12
2	4	1.41	18	324	4.24
3	9	1.73	19	361	4.36
4	16	2	20	400	4.47
5	25	2.24	21	441	4.58
6	36	2.45	22	484	4.69
7	49	2.65	23	529	4.80
8	64	2.83	24	576	4.90
9	81	3	25	625	5
10	100	3.16	26	676	5.10
11	121	3.32	27	729	5.20
12	144	3.46	28	784	5.29
13	169	3.61	29	841	5.40
14	196	3.74	30	900	5.48
15	225	3.87	31	961	5.57
16	256	4	32	1024	5.66

TERMS AND MEANINGS

Many English words used in mathematics have technical meanings:

WORD	MEANING	EXAMPLES
and	addition	4 *and* 6 is 10
difference	subtraction	*difference* of 8 and 6 is 2
of	multiplication	3% *of* 60 = 1.8
per	divide	12 gallons *per* mile
product	result of multiplying	*product* of 4 and 7 is 28
quotient	result of dividing	*quotient* of 6 and 3 is 2
sum	result of adding	the *sum* of 4 and 9 is 13
times	multiplication	7 *times* 5 = 35

UNITS - TABLE OF EQUIVALENTS

SOME USEFUL CONVERSIONS

16 ounces = one pound	365 days = one year
2000 pounds = one ton	366 days = one leap year
32 ounces = one quart	10 years = one decade
4 quarts = one gallon	12 items = one dozen
2 pints = one quart	12 dozen = one gross
55 gallons = one barrel	1000 grams = one kilogram
12 inches = one foot	1000 millirems = one Rem
3 feet = one yard	1 week = seven days
5,280 feet = one mile	28 days = one lunar month
60 seconds = one minute	24 hours = one day
60 minutes = one hour	100 years = one century

Note that these conversion ratios are all equal to ONE. 16 ounces IS one pound, therefore you can multiply any quantity by any of these ratios and not change the value of the quantity.

Note that the ratios work *forwards* and *backwards*, i.e., 16 ounces = one pound and one pound = 16 ounces.

143

VOLUME

SYMBOL V

Volume has three dimensions, length, width and height. It is a measure of the amount of space taken up by an object. Volumes are expressed in cubic units such as cubic feet (ft^3,) cubic centimeters (cc,) quarts, gallons, barrels. The units are *cubed.* Volume is different from *area*. Area has two dimensions, volume has three.

Some volume formulas:

Rectangular solid $V = L \times W \times H$

Cube $\qquad\qquad V = S^3 \qquad$ (S = side)

Sphere $\qquad\quad V = 4/3\ pi \times R^3 \quad$ (R = radius)

Pyramid or Cone $V = 1/3$ area base $\times H$

Cylinder $\qquad\quad V = pi \times R^2 \times H$

An important general idea is that area x distance = volume. 145

ZERO FACTS

Zero is neither negative or positive, but it is an even number because it is located between two odd numbers, - 1 and + 1.

- 5 - 4 - 3 - 2 - 1   0   1   2   3   4   5

Zero divided by any number is zero. 0/3 = 0

Zero multiplied by any number is zero. 0 x 5 = 0

Division by zero is not defined. Division by zero is not permitted.

Any number raised to the zero power is 1.

Example: 5^0 = 1

Zero added to any number leaves that number unchanged.

Example: 6 + 0 = 6

CONTINUED NEXT PAGE   **147**

Zeroes are used as "place-keepers" when the decimal point is moved beyond existing digits in the number.

Example: $0.16 \div 100 = 0.0016$

A leading zero is used in a decimal to alert the reader to look for a decimal point: 0.143 rather than .143 (both are correct, but math used in science prefers 0.143

ZERO POWER OF A NUMBER

Any number raised to the ZERO power is ONE. (Except 0^0 **can't be done.)**

Inspect the following pattern:

$$1000 = 10^3 = 10 \times 10 \times 10$$
$$100 = 10^2 = 10 \times 10$$
$$10 = 10^1 = 10$$
$$1 = 10^0 = 1$$

Here is what is happening. The 1000 is divided by 10 to get 100; the 100 is divided by 10 to get 10; the 10 is divided by 10 to get 1. At the same time, 10^3 is divided by 10 to get 10^2; then divided again by 10 to get 10^1. Each division by ten reduces the exponent by 1. At the point where there are "no factors" of 10, we get 1 by pattern. This pattern will work for any whole number:

$$3^3 = 3 \times 3 \times 3$$
$$3^2 = 3 \times 3$$
$$3^1 = 3$$
$$3^0 = 1$$

149